# YOUR KNOWLEDGE HAS VALUE

- We will publish your bachelor's and master's thesis, essays and papers

- Your own eBook and book - sold worldwide in all relevant shops

- Earn money with each sale

Upload your text at www.GRIN.com and publish for free

**Firdos Ahmad**

# Managing the Forest for Sustainable Development: A Study of Indian States

GRIN Verlag

**Bibliografische Information der Deutschen Nationalbibliothek:**

Die Deutsche Bibliothek verzeichnet diese Publikation in der Deutschen National-
bibliografie; detaillierte bibliografische Daten sind im Internet über http://dnb.d-
nb.de/ abrufbar.



**Imprint:**

Copyright © 2011 GRIN Verlag GmbH
Druck und Bindung: Books on Demand GmbH, Norderstedt Germany
ISBN: 978-3-656-07424-3

**This book at GRIN:**

http://www.grin.com/en/e-book/182917/managing-the-forest-for-sustainable-
development-a-study-of-indian-states

# Managing the Forest for Sustainable Development: A Study of Indian States

By

**Md. Firdos.Ahmad**

Department of Economics,
A.M.U. Aligarh, U.P

## Abstract

*Forest is an important renewable resource and play vital role in the economic development of a developing country like India since the existence of human civilizations. They are rich sources of energy, housing, firewood, timber and fodder and they provide employment to a large section of the rural population. They also plays critical role in maintaining the ecological balance. The present consumption level and supply of forest products reveals that there is shortage of supply of forest products in comparison to its demand for various purposes except for wrapping, packaging paper and paper board and it is also expected to increase in near future because of rising population and growing economy which will make the realization of goal of sustainable development more difficult. Since we can neither stop the growth of population nor we can restrict the development of our country, there is only one option i.e. increasing the forest cover area and the density of existing forests. For this purpose we will find out those states in which there is shortfall of forest cover against the standard forest area. The result of the study reveals that the states like Arunachal Pradesh, Manipur, Meghalya, Mizoram, Assam, Nagaland and Tripura are in comfortable position and there is no need to over emphasis on the management of forest in these states. But for the states like Andhra Pradesh, Bihar, Gujarat, Maharashtra, Punjab, Rajasthan Tamil Nadu,Uttar Pradesh, West Bengal etc. the forest cover area is not comfortable and there is need to manage the forest and adopt the policy of aforestation especially in wasteland.*

**Introduction**

Sustainable development implies use of natural resources such that the future generations can attain the same level of well being as enjoyed by the present generation (WCED1987). It is in context of the need for conservation of the stock of natural resources that sustainable management of forests has gained importance. Forest management deals with the organization of a forest property for its proper maintenance and utilization according to the wishes of the  proprietor (Sagreiya, K.P.1989)

Forest is an important renewable resource and play vital role in the economic development of a developing country like India. Forests are sustenance and survival for a large population. Forests occupied a share of 1.7 percent to gross domestic product (Singh and Beniwal, 1995). Forests fulfill the needs of large section of population by providing the valuable products like timber, fuel wood, and fodder, raw material for many industries and medicinal herbs for the people. Even developed economies have used forests as a major resource due to its utility as wood for homes, furniture, fuel, paper, bark for water proofing material, nuts fruit etc. (Purdom et-al.1980).Further recent researchers have proved that forested lands release water slowly and prevent flood line (N.H.Ravindranath. et.al, 2008). The forest ecosystem maintains the soil eco-system on which our agriculture and food supply is dependent (Shafi, 1992).

---

* Department of Economics, A.M.U, Aligarh

The entire tribal ecology is dependent on forest eco-system. The tribes collect and sell minor forest products such as dry and fallen wood for fuel small timber, nuts, bamboos, hides, skins and herbs etc and these are the main source of income for these people.(Mathur and Soni,1990)

The increased demand for forest products because of fast population growth, urbanization, high rate of economic growth and trade liberalization are putting pressure on forest resources. The demand for food to feed increasing population causing extension in agriculture and shifting cultivation resulted into decline in area under forests. During last two decades, India witnessed annual depletion of forest cover at rate of 253 square kilometer (Anon-1999). To meet this challenge country should improve the management of forest resources, by recognizing the needs of present and future generation. Demand-Supply management implies the management of supply of goods and services in such manner that it increases to a level of required demand. Therefore countries should emphasis on sustainable management of natural forest and expansion of forests through afforestation and farm forestry. Apart from this, country should also focus on management of demand by linking consumption to needs, practicing conservation in the use of forest products, improving the use of fuel wood, promoting efficiency in wood processing industries, promoting efficient pricing of forest products and promoting timber substitutes.

Supply management is also important for sustainable forest management. Forests will continue to be an important source of economic and environmental goods and services. Tree planting in degraded forest resource. Enhancement of productivity in potential forests as well as plantations will be other important areas for an increase in area under forest of up to one third of geographical area as recommended by the

Indian forest policy to meet the requirement of forest products as well as the protection of ecological assets.

. In this situation what we can do is simply managing the forest and forest resources for the minimum and best use. At the same time since forest is a renewable resource we may increase the forest cover by aforestation programme. For this purpose it is important to know that what should be area of forest cover and its growth so as to meet with the demand of rural poor living in and around the forest as well as the demand from construction sector and agro/forest based industries to realize the objective of sustainable development

**Forest Resources and their management:**

The falling forest cover is a victim of rising population. The rise in demand of forest product has been so high because of high growth of population that Indian forest felled to supply it by natural process and depletion of forest cover started. At this juncture it is important to point out that if we want to realize the objective of sustainable development we are bound to manage the forest in such away that will supply the required volume of forest product without making any harm to the ideal level of forest cover. This can be done through assessing the level of demand for all purposes. In general forest products are demanded for two purposes. Firstly, forest products are demanded in huge amount from the urban and rural centres in the construction sector and as a raw material in agro / forest based industries (Maini J.S1991). Secondly, it is demanded by the settlement located in an around the forest for their livelihood. Demand from them is need based. Most of the families that depend on forest products are poor and involved in food gathering, fodder collection, firewood collection and extraction of other forest products like honey etc. Exploitation by these rural poor is not so high that cannot be managed in sustainable manner. The problems of forest management for sustainable development get aggravated when this poor people join hands with the other group for a meager amount of money and help in over extraction of forest resources. The following table gives a detail account of demand of forest product in terms of consumption and the supply for various purposes.

**Table 1**

**Actual Consumption and Supply of Forest Products**

| S. No | | forest products | 1970-72 | | | 1980-82 | | | 1998-00 | | |
|---|---|---|---|---|---|---|---|---|---|---|---|
| | | | actual cons | Actual supply | Imb (%) | actual cons | actual supply | Imb (%) | actual cons | actual supply | Imb (%) |
| 1 | | Round Wood | 178.31 | 178.32 | 0.005 | 228.69 | 228.68 | -0.004 | 284.11 | 282.52 | -0.55 |
| | a | Fuel and Wool Charcoal | 164.95 | 164.95 | | 208.86 | 208.85 | -0.004 | 261.14 | 259.79 | -0.1 |
| | b | industrial round wood | 13.36 | 13.38 | 0.14 | 19.83 | 19.83 | 0 | 22.97 | 22.73 | -1.04 |
| 2 | | Wood based products (Million M3) | 0.17 | 0.18 | 5.88 | 0.22 | 0.23 | -4.54 | 0.27 | 0.29 | 7.4 |
| | a | ply based wood | 0.03 | 0.14 | 0.366 | 0.17 | 0.18 | -5.88 | 0.04 | 0.08 | 100 |
| | b | Partical board | 0.01 | 0.01 | 0 | 0.02 | 0.02 | 0 | 0.09 | 0.04 | 0 |
| | c | Fiber board | 0.03 | 0.03 | 0 | 0.03 | 0.03 | 0 | 0.14 | 0.13 | 7.14 |
| 3 | | Sawn Wood (million M3) | 4.76 | 4.75 | -0.21 | 10.99 | 10.98 | 0 | 16.29 | 16.29 | 0 |
| 4 | | Wood Pulp (Million M3) | 0 | 0.15 | -25 | 0.58 | 0.48 | -17.24 | 1.75 | 1.55 | 11.42 |
| 5 | | paper and paper board (million tonnes) | 1.02 | 0.84 | -18 | 1.98 | 1.167 | -15.65 | 3.86 | 3.61 | -6.47 |
| | a | news print | 0.21 | 0.04 | -80.95 | 0.39 | 0.11 | -71.79 | 0.61 | 0.44 | -27.86 |
| | b | printing and writiing paper | 0.45 | 0.45 | 0 | 1.01 | 0.99 | -1.98 | 1.49 | 1.34 | -10.06 |
| | c | other paper and paper board | 0.36 | 0.35 | -2.77 | 0.58 | 0.56 | -3.44 | 1.75 | 1.73 | -1.14 |
| 6 | | Wrapping and packaging paper and paper board (Million tones) | 0.15 | 0.14 | -6.66 | 0.45 | 0.45 | 0 | 1.13 | 1.13 | 0 |

Where: Imb: imbalances, Cons: Consumptions,

Source: Malik.D.P.and Sunil Dhanda (2003); Status, Trends and Demand for Forest Products in India, Table.3 and Table.4.

It can be observed from the above table that in 1970-72 there were marginal imbalances in the consumption and supply of the forest product. The items in which we have a surplus supply are round wood, industrial round wood, wood based panels and ply based wood and the items in which there is deficiency in supply are swan wood, wood pulp, paper and paper band, news print, wrapping and packaging paper and paper board. The situation is quite different in 1998-00. In the most of the items of forest products there is a short fall of supply except a few items. The items in which there is remarkable increase in the demand are round wood, fuel wood, charcoal, wood pulp, paper and paper board, news and printing and writing paper. But as far as the supply of these items is concern the increase is not as much as in the demand.

**Table 2**

**Projected Consumption and Supply of Forest Products**

| S. No | | forest products | 2005 | | | 2010 | | |
|---|---|---|---|---|---|---|---|---|
| | | | Projected cons | Projected supply | Imb (%) | Projected cons | Projected supply | Imb (%) |
| 1 | | Round Wood | 342.06 | 329.51 | -3.67 | 376.43 | 366.48 | -2.64 |
| | A | Fuel and Wool Charcoal | 304.98 | 299.4 | -1.82 | 334.51 | 331.23 | -.98 |
| | B | Industrial round wood | 37.08 | 30.11 | -18.79 | 41.92 | 34.25 | -18.29 |
| 2 | | Wood based products (Million M3) | .38 | .30 | -10.52 | 0.54 | .45 | -16.67 |
| | A | ply based wood | .07 | .05 | -28.57 | 0.11 | .08 | -27.27 |
| | B | Partical board | .15 | .12 | -20.0 | 0.23 | .15 | -34.78 |
| | C | Fiber board | .16 | .17 | 6.25 | 0.20 | .22 | 10.00 |
| 3 | | Sawn Wood (million M3) | 45.86 | 30.84 | -32.75 | 66.17 | 37.23 | -43.73 |
| 4 | | Wood Pulp (Million M3) | 3.36 | 2.94 | -12.5 | 5.18 | 4.68 | -9.65 |
| 5 | | paper and paper board (million tonnes) | 6.08 | 6.13 | -0.82 | 8.57 | 8.85 | 3.62 |
| | A | news print | 1.82 | 1.93 | 8.24 | 3.07 | 3.45 | 13.02 |
| | B | printing and writiing paper | 1.75 | .97 | -44.57 | 2.14 | 1.04 | -51.4 |
| | C | other paper and paper board | 2.51 | 3.23 | 28.68 | 3.36 | 4.36 | 29.76 |
| 6 | | Wrapping and packaging paper and paper board (Million tones) | 4.11 | 4.19 | 1.96 | 7.0 | 7.15 | 2.42 |

Where: Imb: imbalances, Cons: Consumptions,

Source: Same as Table 1

Note: Projection was made on data available for 1970-1994.

The above table represents the projected consumption and supply of forest products in 2005 and 2010. The high growth of population and the development process are expected to increase the pace of growth in projected demand of forest products however on the other hand the diversion of forest land for other purposes may cause a decline in the supply of forest products. The difference in consumption and supply of forest products though projected to occur in most of the products but a wide gap may be in case of round wood, industrial round wood, wood based panel, ply based wood, particle board, swan wood and printing and writing paper.

**State-wise Imbalances of Forest Resource**

The above discussion about the consumption level and supply of forest products reveals that there is shortage of supply of forest products in comparison to its demand for various purposes and it is also expected to increase in near future because of rising population and growing economy which will make the realization of goal of sustainable development more difficult. Since we can neither stop the growth of population nor we will restrict the development of our country, there is only one option i.e. increasing the forest cover area and the density of existing forests. Moreover the problem of finance is another obstacle in the way of aforestation/reforestation along with the availability of land. To make this process of aforestation comfortable it is necessary to concentrate in some specified area rather than the whole country. For this purpose we will find out those states in which there is shortfall of forest cover against the standard forest area. A detail is given in the following table.

## Table 3

**States with Forest Cover more than 33 per cent as Percentage of Gross Area**

| States | 1971 | 1991 | 2001 | 2005 |
| --- | --- | --- | --- | --- |
| Arunachal Pradesh | 61.5 | 81.8 | 81.25 | 80.93 |
| Assam | 26.87 | 33.1 | 35.33 | 35.24 |
| Kerala | 22.12 | 26.1 | 40.03 | 40.13 |
| Manipur | 67.53 | 80.1 | 75.85 | 76.53 |
| Meghalya | 64.03 | 70.98 | 69.48 | 75.74 |
| Mizoram | 65.91 | 89.47 | 82.98 | 88.63 |
| Nagaland | 49.61 | 86.12 | 80.49 | 82.75 |
| Tripura | 60.11 | 50.78 | 67.37 | 77.77 |

**States with Forest Cover less than 33 per cent as Percentage of Gross Area**

| | | | | |
| --- | --- | --- | --- | --- |
| Andhra Pradesh | 17.7 | 17.40 | 16.22 | 16.13 |
| Bihar | 13.05 | 15.5 | 6.07 | 5.92 |
| Gujrat | 4.85 | 5.9 | 7.72 | 7.51 |
| Haryana | 1.81 | 1.27 | 3.96 | 3.59 |
| Himachal Pradesh | 27.12 | 24.0 | 25.79 | 25.81 |
| Jammu & Kashmir | 10.03 | 9.2 | 9.55 | 9.57 |
| Karnataka | 15.38 | 16.8 | 19.28 | 18.38 |
| Madhya Pradesh | 24.52 | 30.03 | 25.06 | 24.66 |
| Maharashtra | 13.22 | 14.32 | 15.43 | 15.43 |
| Orissa | 31.07 | 30.26 | 24.94 | 31.09 |
| Punjab | 2.18 | 2.32 | 4.82 | 3.09 |
| Rajasthan | 3.33 | 3.80 | 4.78 | 4.63 |
| Tamil Nadu | 12.84 | 13.62 | 16.51 | 17.72 |
| Uttar Pradesh | 8.80 | 11.49 | 5.7 | 5.86 |
| West Bengal | 9.45 | 9.46 | 12.18 | 13.99 |
| **India** | 16.89 | 19.49 | 20.05 | 20.60 |

Source 1; Forest Survey of India (2009); Ministry of Environment and Forest,GOI,
Dehradun.

The above table provides information about the forest cover in Indian states in percentage term. The table is divided into two parts. First part covers states with forest cover more than 33.00 percent of gross area and the second part represents state with low forest cover. The states with high forest cover area have already attained the required forest cover and there is no need to worry about the forest cover to attain the goal of sustainable development. The states like Arunachal Pradesh, Assam, Kerala, Manipur, Meghalya, Mizoram, Nagaland and Tripura are in comfortable position and there is no need to overemphasis on the management of forest in these states. But for the states likes Andhra Pradesh, Bihar, Gujarat, Haryana, Himachal Pradesh, Jammu&Kashmir, kanataka, Madhya Pradesh, Maharashtra, Orissa, Punjab, Rajasthan and Tamil Nadu the forest cover area is not comfortable and there is need to manage the forest and adopt the policy of a forestation.

The other part of the table represents the states with low forest cover. In Andhra Pradesh the forest cover area decreased from 17.7 percent in 1971 to 16.13 percent in 2005. In Bihar also the forest cover area decreased from 13.05 percent to 5.92 percent in 2005. In Gujrat there has been a slight increase in the forest cover from 4.85 percent in 1971 to 7.51 percent in 2005. In case of Haryana though the forest cover area has increased over the years but still it comes in lowest forest cover states. A decline in the forest cover of Himachal Pradesh has brought it below the boundary line of required forest cover. Jammu & Kashmir has also witnessed a slight fall in the forest cover area. Against this states like Madhya Pradesh, Maharashtra, Orissa, Tamil Nadu and Karnataka have observed some rise in the forest cover area and marched towards required forest cover so as to maintain the environmental quality. It can be therefore said that these are the states with greater potential for increasing the forest cover area and make the process of development sustainable.

## Wasteland utilization—

Since the supply of forest product is constrained by forest area hence concentrated efforts are required to raise the forest cover to the level of one third of geographical area (Indian forest policy) and it has become an urgent economic activity. Table 4 gives an account of these states where forest area is less than one third of geographical area along with wasteland area.

**Table.4**
**Forest Cover and Wasteland in India**

| States | Forest Area (in percentage) | Wasteland (in percentage) |
|---|---|---|
| Andhra Pradesh | 16.13 | 18.81 |
| Bihar | 05.92 | 12.08 |
| Gujarat | 7.51 | 21.95 |
| Haryana | 3.59 | 08.45 |
| Himachal Pradesh | 25.81 | 56.87 |
| Jammu & Kashmir | 09.57 | 64.55 |
| Karnataka | 18.38 | 10.87 |
| Madhya Pradesh | 24.66 | 15.72 |
| Maharashtra | 15.43 | 17.38 |
| Orissa | 31.09 | 13.71 |
| Punjab | 03.09 | 04.42 |
| Rajasthan | 04.63 | 30.87 |
| Tamil Nadu | 17.72 | 12.17 |
| Uttar Pradesh | 05.86 | 13.17 |
| West Bengal | 13.95 | 06.44 |
| India | 20.60 | 20.17 |

Source: Forest Survey of India (2009), Ministry of Environment, Government of India,

Dehradun

Above mentioned states can be divided into three categories. In first category we included those states in which the desired level of forest area can easily managed by using some part of wasteland. States like Himachal Pradesh, Jammu and Kashmir, Orissa and Rajasthan may be included. The second category includes in this category those states where serious effort is required to attain the desired level of forest cover area. This category includes the states like Andhra Pradesh, Maharashtra and Tamil Nadu. In the last category we include Gujarat, Bihar, Haryana, Karnataka, Punjab, Uttar Pradesh and west Bengal in which the desired level of forest cover area can't be managed even by using whole wasteland. Therefore, alternative policy is needed to attain the desired level of forest cover area either by increasing the plant intensity or by expending the policy of social forestry even in residential areas by adopting the concept of green building

**Conclusion-**

India is a developing country with a high population density and low forest area per capita. The livestock population density is among the highest in the world. Further nearly 70 percent of the population residing in rural areas depends on forest and other biomass resource for its energy needs and livelihood. Today forest is considered more than the resource and due to degrading environment resulting from deforestation, there is a greater concern for consideration and preservation of this biotic resource. Thus, there is a need for an increase in area under forest at least up to one third of total geographical area as recommended by the Indian forest policy to meet the requirement of forest products as well as the production of ecological assets. From the above analysis, it may concluded that the consumption level and supply of forest products reveals that there is shortage of supply of forest products in comparison to its demand for various purposes and it also expected to increase in near future because of rising population and growing economy which will make the realization of goal of sustainable development more difficult. The items in which there is remarkable increase in the demand are round wood, fuel wood, charcoal, wood pulp, paper and paper board news and printing and writing paper, but as far as the supply of these items is concern the increase is not as much as in the demand. Further, the difference in consumption and supply of forest products though projected to occur in most of the products but a wide gap may be in case of round wood, industrial round wood, wood based panel, ply based wood, particle board, swan wood and printing and writing.

Due to limited resources finance has become an obstacle in the way of aforestation programme. To make this process of aforestation comfortable it is necessary to concentrate in some specified area rather than concentrating on whole country. For this purpose we find out states viz,Karnataka, Madhya Pradesh, Maharashtra, Orissa Punjab, Rajasthan, Tamil Nadu, Uttar Pradesh and west Bengal in which there is shortfall of forest against the standard forest area. Efforts should be made to increase the area under forests cover by aforestating wasteland through social and agro-forestry by way of people's participation. Therefore, some states like Himachal Pradesh, Jammu Kashmir, Madhya Pradesh, Orissa, Rajasthan, Andhra Pradesh, Maharashtra and Tamil Nadu are identified in which use of some part of wasteland can help to attained desired level of forest cover. Sustainable livelihoods in natural forest regions will require the participation of the community. The state will have to intervene to secure the basic needs of the people living in forest regions.

**References**

1. Annon (1999), *National Forestry Action Programme*-India, Ministry of Environment and Forest, GOI, New Delhi.

2. N.H.Ravindranath, et.al (2008), '*Forest Conservation, Afforestation and Reforestation in India: Implications for Forest Carbon Stocks*', Current Science. Bangalore, India.

3.Mathur H.N and Soni.P(1990), Forests;Their Role in Present Day Life, in Gupta K.M.(ed) *Himalayas Man and Nature*, Lancers Books, New Delhi.

4. Maini.J.S.(1991) *"Guiding principles towards a global concerns for the conservation and sustainable development of all types of forests world wide"*, attawa,Canada

5. Sagreiya,K.P.(1989) *"Forest and Forestry"*, National Book Trust, New Dehli, India.

6. Shafi. M.(1992 ) Utilisation and conservation of forests in India with special reference to social forestry in Shafi.M. and Mehdi.R,(ed) *Forest ecosystems of the world*, Rawat Publication, New Dehli.

7. Singh,K. and S.K. Beniwal (1995), *Socio-Economic development through community forestry.Seminar on community forestry bio diversity*, ISTL,Solan,India.

8. Purdom P.W and Anderson S.H(1980). *Environmental Science managing the environment and forest,* GOI, New Dehli.

9. World Commission on Environment and Development (WCED). 1987. Our Common   Future. Oxford, Oxford University Press.